OUR FAVORITE BRANDS

YOUTUBE

By Emma Huddleston

Kaleidoscope
Minneapolis, MN

The Quest for Discovery Never Ends

This edition is co-published by agreement between Kaleidoscope and World Book, Inc.

Kaleidoscope Publishing, Inc.
6012 Blue Circle Drive
Minnetonka, MN 55343 U.S.A.

World Book, Inc.
180 North LaSalle St., Suite 900
Chicago IL 60601 U.S.A.

All rights reserved. No part of this book may be reproduced in any form without written permission from the publishers.

Kaleidoscope ISBNs
978-1-64519-023-3 (library bound)
978-1-64494-188-1 (paperback)
978-1-64519-123-0 (ebook)

World Book ISBN
978-0-7166-4323-4 (library bound)

Library of Congress Control Number
2019939236

Text copyright ©2020 by Kaleidoscope Publishing, Inc. All-Star Sports, Bigfoot Books, and associated logos are trademarks and/or registered trademarks of Kaleidoscope Publishing, Inc.

Printed in the United States of America.

FIND ME IF YOU CAN!

Bigfoot lurks within one of the images in this book. It's up to you to find him!

TABLE OF
CONTENTS

Chapter 1: Sharing Videos **4**

Chapter 2: History of YouTube **8**

Chapter 3: Create, Watch, Listen, Go **16**

Chapter 4: YouTube Around the World **22**

    Beyond the Book 28

    Research Ninja 29

    Further Resources 30

    Glossary 31

    Index 32

    Photo Credits 32

    About the Author 32

CHAPTER 1

FUN FACT
YouTube did a livestream of a 2019 music festival called Coachella. The stream got more than 82 million live views.

People of all ages enjoy watching YouTube.

Sharing Videos

Molly started watching YouTube in 2009. She saw a **viral** video. It was called "Charlie Bit My Finger—Again!" Baby Charlie bit his older brother's finger. He wouldn't let go. Charlie's older brother cried out. He made a funny face. Molly laughed. The video had 129 million views. That made it the most-watched video on YouTube. Molly showed her friends. YouTube made video sharing easy. Before YouTube, Molly had to email video clips. But it was hard. Most video files were too big.

Soon, Molly was watching lots of **content**. She saw movie trailers. She watched music videos. YouTube also connected with television networks. Molly watched a livestream of the 2012 Olympics. A livestream is a video recording that is shared online in real time.

Many people like to document their travels through pictures or videos.

Ty hops on a city bus. He pulls out his phone. Ty has the YouTube Studio app on his phone. He logs in. His account lets him **upload** videos. People without accounts have time limits. Their videos can't be more than fifteen minutes long. But Ty's account gives him more time. His videos can be any length. Ty starts filming himself. He has a YouTube channel. It is about travel. He is visiting London today. The bus drives down the streets. Stone buildings blur together outside the window.

Someone gives Ty's video a "like." The number of views on his video grows. People leave comments. They give ideas of places to visit. Ty smiles. He gets off at the next stop. Exploring London is even more fun with YouTube. Ty can share his adventures with others.

FUN FACT
In 2019, YouTube said that 70 percent of all its videos were seen on a mobile device.

People can use YouTube on many devices, such as phones and laptops.

CHAPTER 2

History of YouTube

It was early in the 2000s. Three friends had a problem. Chad Hurley, Steve Chen, and Jawed Karim worked with computers. They shared photos online. They wanted to share videos, too. But there was no easy way to do that.

In the past, Chad Hurley, left, and Steve Chen have spoken to reporters about YouTube.

FUN FACT
In 2019, Hurley and Chen won a Lifetime Achievement Emmy award.

The friends worked together. They made a website. It would let people share videos. YouTube went online in February 2005. The friends posted the first video in April. It was titled "Me at the Zoo." Karim went to the San Diego Zoo in the video. He visited elephants. The video was eighteen seconds long.

YouTube released its beta version in May. A beta version of a product is not quite complete. It is being tested before the product's official release. It was a hit. More than 30,000 people visited it each day. YouTube officially **launched** in December 2005.

YouTube grew quickly. It had millions of visitors each day. Companies supported the founders with money. This helped the founders keep up with the number of visitors. YouTube reached a milestone in the summer of 2006. It had more than 100 million views a day. The site had barely been up for a year.

Today, people can watch news clips from different countries on YouTube.

Google has a YouTube office in California.

Google bought YouTube in November 2006. It paid $1.65 billion. Google had more resources. It could help with legal issues. It could advertise the site. Hurley and Chen still kept some control of YouTube. They didn't want ads to take over. Videos were still free to watch.

Hurley and Chen kept updating YouTube. It's one of the most popular websites in the world today. In 2017, 26 percent of US internet users visited YouTube multiple times per day. Only 16 percent of US internet users said they never used YouTube.

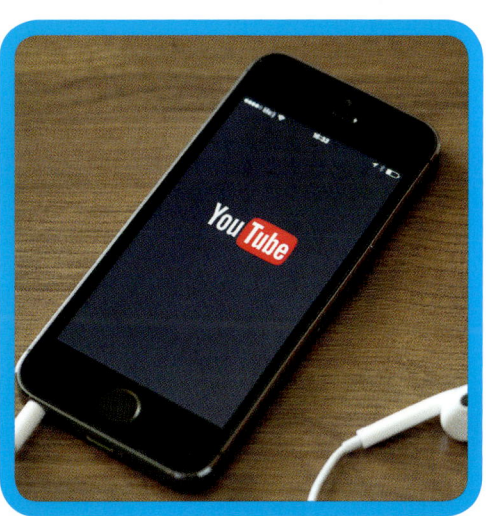

YOUTUBE TIMELINE

April 23, 2005
The first video, "Me at the Zoo," is posted to YouTube.

December 15, 2005
YouTube officially launches.

November 13, 2006
Google acquires YouTube for $1.65 billion.

2005 — 2006 — 2007 — 2008 — 2009 — 2010 — 2011

September 2005
A Nike ad becomes the first video to reach one million views.

April 2009
YouTube teams up with Vevo to make a music video streaming service.

May 2005
The beta version of YouTube launches.

February 14, 2005
Hurley, Chen, and Karim start YouTube.

October 2015
YouTube launches YouTube Red. The paid service allows customers to watch without ads, download videos, and more.

May 2018
YouTube Red becomes YouTube Premium.

2017
YouTube ends its paid channel service.

May 2013
YouTube starts a paid channel service. Some channels can charge users to watch their videos.

2012 2013 2014 2015 2016 2017 2018

December 2012
The viral music video for "Gangnam Style" becomes the first video to reach one billion views.

2016
YouTube begins producing its own original series.

November 2018
YouTube Stories becomes available to creators with 10,000 or more subscribers.

May 2019
YouTube reaches two billion users per month.

CHAPTER 3

Create, Watch, Listen, Go

Aisha dances in her room. Music plays from her laptop. She follows Taylor Swift's YouTube channel. Aisha has a YouTube Premium **subscription**. She watches music videos. Her friends do, too. But she doesn't have to watch ads.

Aisha got YouTube Premium in 2018. It cost $12 a month. She thinks it's worth it. Her music isn't interrupted by ads. And she can download music videos.

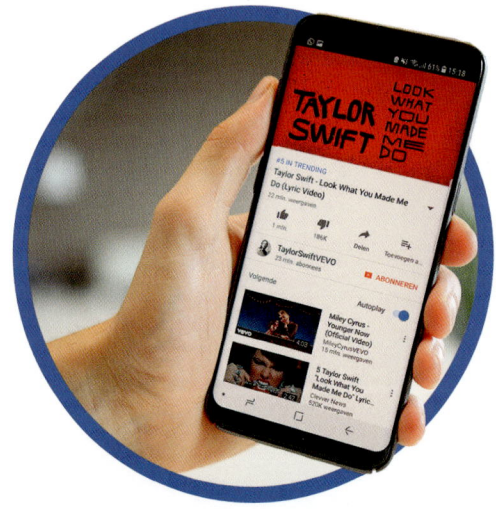

On YouTube, people can experience different kinds of music.

Justin Bieber's YouTube videos racked up millions of views.

YouTube changed the music world. It let people watch music videos any time. Famous singers were discovered on YouTube. They posted videos of themselves singing. Justin Bieber did that in 2007. YouTube helped make him a star. Aisha still watches him on YouTube. But now she watches his official music videos.

Sometimes she uses captions on music videos. They help her learn lyrics. YouTube has automatic captions. The captions are in ten languages. A computer program listens to videos. It can tell what is being said. Then it shows the words on the screen. These captions help people who can't hear. They let more people enjoy YouTube.

Liam liked watching sports. But his family didn't have cable. He had to go to his friend's house to see big games. In 2017, YouTube TV came out. It cost $35 a month. He asked his mom about getting it. She could watch the news. His little sister could watch cartoons. And Liam could watch sports on ESPN.

FUN FACT
In 2019, YouTube found out that more 18- to 49-year-olds watch YouTube than watch cable TV networks.

Liam's mom thought it was a great idea. She signed the family up. Everyone watches their favorite channels. They can search for specific football games. They can watch their favorite TV episodes. Regular TV channels only show one show at a time. YouTube lets Liam view anything at any time. And it works on any device with internet access.

Ray browses YouTube. He goes to the "Stories" page. YouTube added a Stories feature in 2018.

People with more than 10,000 subscribers could use it. Ray follows a few popular channels. The creators livestream while playing video games. They have millions of followers.

Ray watches one creator's story. It's made up of short video clips. Some have stickers and filters. Ray sees his favorite YouTube creators living their daily lives. He feels connected to them.

Susan Wojcicki, YouTube's CEO, introduced YouTube TV on February 28, 2017.

CHAPTER 4

YouTube Around the World

Marco watches movies and TV. But he sees they mostly have only white people. He has dark skin. He sometimes feels left out. He wants to see more people of color on-screen. YouTube is different. Marco thinks it gives **minorities** a voice. Anyone can post. Anyone can share their story. YouTube is available in eighty languages and ninety-one countries.

Some popular YouTube channels review different types of make-up.

People make tutorial videos that they hope will help others.

Marco uses YouTube for news. In 2017, Hurricane Harvey hit Texas. He found a video about it on YouTube. He watched it for details. And he watched political videos in 2018. It was his first time voting. He wanted to be informed. YouTube helped him learn about the issues. But he doesn't only watch serious content. He likes videos that teach him new skills. He is learning guitar.

YouTube changed the way Mia learns. She goes online for homework help. Khan Academy has a YouTube channel. It makes video lessons. Mia watches math and history lessons. The videos cover many other subjects, too. Mia likes learning from videos. She can pause or rewind. Other people like learning on YouTube, too.

YOUTUBE PARTNER PROGRAM

YouTube offers a Partner Program. This lets creators make money. Creators play ads. Then they are paid for ads viewed. A creator named Ryan makes money this way. Ryan reviews toys for other kids. His channel has 17 million subscribers. He made $22 million in 2018. He was seven years old.

It was 2010. Dan Savage and Terry Miller made a video. They wanted to help **LGBTQ+** kids and teens. This group of people is often bullied. Savage and Miller sent a message of hope. Their video made a big impact. They started the It Gets Better Project. People upload messages. They give encouragement. They tell young LGBTQ+ viewers that things will get better. It Gets Better has its own YouTube channel.

The It Gets Better Project started with one video. YouTube makes sharing content easy. One video can change the world.

VidCon is a YouTube convention. YouTube creators and fans gather together.

YOUTUBE VIEWS
Per Day

- **Dec. 2005** 2 million
- **Jan. 2006** 25 million
- **July 2006** More than 100 million
- **July 2008** More than 200 million
- **May 2010** More than 2 billion
- **Aug. 2012** More than 4 billion
- **Dec. 2017** More than 5 billion
- **May 2019** More than 1.9 billion users each month, 1 billion hours of content viewed each day

BEYOND THE BOOK

After reading the book, it's time to think about what you learned. Try the following exercises to jumpstart your ideas.

THINK

THAT'S NEWS TO ME. The first video posted on YouTube was "Me at the Zoo." Consider how news sources might be able to fill in more detail on the beginnings of YouTube. What new information could be found in news articles? Where could you go to find those sources?

CREATE

SHARPEN YOUR RESEARCH SKILLS. YouTube grew quickly. Where could you go in the library, or who could you talk with, to find more information about YouTube's fast growth? Create a research plan and write a paragraph that details these next steps for research.

SHARE

SUM IT UP. Write one paragraph summarizing the important points from the whole book. Then share the paragraph with a classmate. Does he or she have any feedback on your summary or additional questions about YouTube?

GROW

DRAWING CONNECTIONS. Create a diagram that shows and explains connections between YouTube and computer technology. How does learning about technology help you better understand YouTube?

RESEARCH NINJA

Visit *www.ninjaresearcher.com/0233* to learn how to take your research skills and book report writing to the next level!

RESEARCH

DIGITAL LITERACY TOOLS

SEARCH LIKE A PRO
Learn about how to use search engines to find useful websites.

FACT OR FAKE?
Discover how you can tell a trusted website from an untrustworthy resource.

TEXT DETECTIVE
Explore how to zero in on the information you need most.

SHOW YOUR WORK
Research responsibly—learn how to cite sources.

WRITE

GET TO THE POINT
Learn how to express your main ideas.

PLAN OF ATTACK
Learn prewriting exercises and create an outline.

DOWNLOADABLE REPORT FORMS

Further Resources

BOOKS

Bernhardt, Carolyn. *Film It! YouTube Projects for the Real World.* Abdo, 2017.

Smibert, Angie. *12 Great Moments that Changed Internet History.* 12 Story Library, 2015.

Wooster, Patricia. *YouTube Founders Steve Chen, Chad Hurley, and Jawed Karim.* Lerner, 2014.

WEBSITES

Factsurfer.com gives you a safe, fun way to find more information.

1. Go to www.factsurfer.com.
2. Enter "YouTube" into the search box and click 🔍.
3. Select your book cover to see a list of related websites.

Glossary

content: A website's content is its published videos, music, pictures, and writing. YouTube has a range of content, including homemade videos, livestreams, and news clips.

launched: A product or website is launched when it becomes available for the public to use or buy. YouTube officially launched in 2005.

LGBTQ+: LGBTQ+ is an acronym that stands for Lesbian, Gay, Bisexual, Transgender, Queer/Questioning, and more. It describes gender identities and sexual orientations. The It Gets Better Project provides young LGBTQ+ people with encouraging, supportive videos.

minorities: Minorities are people who have one or more characteristics, such as race, religion, or sexual orientation, that is different from a large part of the population. People who are minorities can find welcoming communities on YouTube.

subscription: A subscription is an ongoing paid service that companies provide to customers. YouTube Premium is a subscription that gives people access to features like ad-free content.

upload: To upload is to transfer data or files from a computer or device to the internet. People upload content to YouTube every day.

viral: Online viral content spreads quickly and widely on the internet. Many videos from YouTube, such as "Charlie Bit My Finger—Again!" have gone viral.

Index

awards, 9

captions, 19
Chen, Steve, 8–9, 13, 14
Coachella, 4

Google, 13, 14

Hurley, Chad, 8–9, 13, 14

It Gets Better Project, 26

Karim, Jawed, 8–9, 14
Khan Academy, 25

languages, 19, 22
livestreams, 4, 5, 21

Partner Program, 25

Stories, 15, 20–21

Vevo, 14
views, 5, 7, 10, 14, 15, 27
viral videos, 5, 15

YouTube Premium, 15, 16
YouTube Studio, 6
YouTube TV, 19–20

PHOTO CREDITS

The images in this book are reproduced through the courtesy of: Veja/Shutterstock Images, front cover (boys); milana_babic/Shutterstock Images, front cover (play icon); Africa Studio/Shutterstock Images, front cover (dog), pp. 16–17; maxim ibragimov/Shutterstock Images, front cover (girl); guruXOX/Shutterstock Images, front cover (skateboarder); GE_4530/Shutterstock Images, front cover (tablets); Alex Segre/Shutterstock Images, front cover (background); Asier Romero/Shutterstock Images, front cover (meme); Alexey Boldin/Shutterstock Images, pp. 3, 14 (top); bbernard/Shutterstock Images, pp. 4–5; Margaret.W/Shutterstock Images, p. 6; NiP STUDIO/Shutterstock Images, p. 7; Christoph Dernbach/picture-alliance/dpa/AP Images, pp. 8–9, 14 (bottom); v777999/iStockphoto, p. 10; John Stillwell/PA Wire URN: 22962471/Press Association/AP Images, pp. 10–11; Asif Islam/Shutterstock Images, pp. 12–13; pressureUA/iStockphoto, p. 13; Anton Gvozdikov/Shutterstock Images, p. 14 (middle); Red Line Editorial, p. 27 (chart); Denys Prykhodov/Shutterstock Images, p. 15 (top); Dragon Images/Shutterstock Images, p. 15 (bottom); ThomasDeco/Shutterstock Images, pp. 16, 19; Debby Wong/Shutterstock Images, p. 18; Reed Saxon/AP Images, pp. 20–21; fiz_zero/Shutterstock Images, p. 22; Artem Varnitsin/Shutterstock Images, p. 23; Peera_stockfoto/Shutterstock Images, p. 24; FrameStockFootages/Shutterstock Images, p. 25; Randy Miramontez/Shutterstock Images, p. 26; Twinsterphoto/Shutterstock Images, p. 27 (view counter); Rose Carson/Shutterstock Images, p. 30.

ABOUT THE AUTHOR

Emma Huddleston enjoys reading and swing dancing. She lives in the Twin Cities with her husband.